THIS BOOK BELONGS TO

EMAIL: _______________________

ADDRESS: _______________________

CONTACT: _______________________

PHONE: _______________________

START DATE	END DATE

MO TU WE TH FR SA SU
☐   ☐   ☐   ☐   ☐   ☐   ☐

DATE: /   /

PROJECT:

FOREMAN:

WEATHER   F°_____ C°_____   _____ AM _____ PM

HOURS DUE TO BAD WEATHER

ISSUED AND DELAYS

NOTE: ____

COMPLETION DATE	DAYS AHEAD OF SCHEDULE	DAYS BEHIND SCHEDULE

SAFETY AND INCIDENTS

SAFETY ISSUES THAT NEED TO BE ADDRESSED	ACCIDENTS / INCIDENTS / STEPS NEEDED TO RESOLVE

SUMMARY OF THE WORK DONE TODAY

IMPORTANT NOTES

NAME	SIGNATURE

TODAY LABOR

INITIALS	TRADE	START	FINISH	PAID HOURS	OVERTIME	COMPANY
☐ EMPLOYEE ☐ CONTRUCTOR		AM	PM			
☐ EMPLOYEE ☐ CONTRUCTOR		AM	PM			
☐ EMPLOYEE ☐ CONTRUCTOR		AM	PM			
☐ EMPLOYEE ☐ CONTRUCTOR		AM	PM			
☐ EMPLOYEE ☐ CONTRUCTOR		AM	PM			
☐ EMPLOYEE ☐ CONTRUCTOR		AM	PM			
☐ EMPLOYEE ☐ CONTRUCTOR		AM	PM			
☐ EMPLOYEE ☐ CONTRUCTOR		AM	PM			

EQUIPMENT ON SITE	NO. OF UNITE	WORKING YES / NO

HIRED EQUIPMENT	NO. OF UNITE	EQUIPMENT RENTED	FROM	RATE

NAME: _______________   SIGNATURE: _______________

MO TU WE TH FR SA SU
☐   ☐   ☐   ☐   ☐   ☐   ☐

DATE: /   /

PROJECT:

FOREMAN:

WEATHER

F°____ C°____ ____ AM ____ PM

HOURS DUE TO BAD WEATHER

ISSUED AND DELAYS

NOTE: ____

COMPLETION DATE	DAYS AHEAD OF SCHEDULE	DAYS BEHIND SCHEDULE

SAFETY AND INCIDENTS

SAFETY ISSUES THAT NEED TO BE ADDRESSED	ACCIDENTS / INCIDENTS / STEPS NEEDED TO RESOLVE

SUMMARY OF THE WORK DONE TODAY

IMPORTANT NOTES

NAME	SIGNATURE

TODAY LABOR

INITIALS	TRADE	START	FINISH	PAID HOURS	OVERTIME	COMPANY
☐ EMPLOYEE ☐ CONTRUCTOR		AM	PM			
☐ EMPLOYEE ☐ CONTRUCTOR		AM	PM			
☐ EMPLOYEE ☐ CONTRUCTOR		AM	PM			
☐ EMPLOYEE ☐ CONTRUCTOR		AM	PM			
☐ EMPLOYEE ☐ CONTRUCTOR		AM	PM			
☐ EMPLOYEE ☐ CONTRUCTOR		AM	PM			
☐ EMPLOYEE ☐ CONTRUCTOR		AM	PM			
☐ EMPLOYEE ☐ CONTRUCTOR		AM	PM			

EQUIPMENT ON SITE	NO. OF UNITE	WORKING YES / NO

HIRED EQUIPMENT	NO. OF UNITE	EQUIPMENT RENTED	FROM	RATE

NAME: _______________________   SIGNATURE: _______________________

<table>
<tr><td>MO</td><td>TU</td><td>WE</td><td>TH</td><td>FR</td><td>SA</td><td>SU</td></tr>
<tr><td>☐</td><td>☐</td><td>☐</td><td>☐</td><td>☐</td><td>☐</td><td>☐</td></tr>
</table>

DATE: ___ / ___ / ___

PROJECT:

FOREMAN:

WEATHER F° ___ C° ___ ___ AM ___ PM

HOURS DUE TO BAD WEATHER	ISSUED AND DELAYS

NOTE: _______________________________________

COMPLETION DATE	DAYS AHEAD OF SCHEDULE	DAYS BEHIND SCHEDULE

SAFETY AND INCIDENTS

SAFETY ISSUES THAT NEED TO BE ADDRESSED	ACCIDENTS / INCIDENTS / STEPS NEEDED TO RESOLVE

SUMMARY OF THE WORK DONE TODAY

IMPORTANT NOTES

NAME	SIGNATURE

TODAY LABOR

INITIALS	TRADE	START	FINISH	PAID HOURS	OVERTIME	COMPANY
☐ EMPLOYEE ☐ CONTRUCTOR			AM | PM			
☐ EMPLOYEE ☐ CONTRUCTOR			AM | PM			
☐ EMPLOYEE ☐ CONTRUCTOR			AM | PM			
☐ EMPLOYEE ☐ CONTRUCTOR			AM | PM			
☐ EMPLOYEE ☐ CONTRUCTOR			AM | PM			
☐ EMPLOYEE ☐ CONTRUCTOR			AM | PM			
☐ EMPLOYEE ☐ CONTRUCTOR			AM | PM			
☐ EMPLOYEE ☐ CONTRUCTOR			AM | PM			

EQUIPMENT ON SITE	NO. OF UNITE	WORKING YES / NO

HIRED EQUIPMENT	NO. OF UNITE	EQUIPMENT RENTED	FROM	RATE

NAME: _______________________   SIGNATURE: _______________________

TODAY LABOR

INITIALS	TRADE	START	FINISH	PAID HOURS	OVERTIME	COMPANY
☐ EMPLOYEE ☐ CONTRUCTOR		AM	PM			
☐ EMPLOYEE ☐ CONTRUCTOR		AM	PM			
☐ EMPLOYEE ☐ CONTRUCTOR		AM	PM			
☐ EMPLOYEE ☐ CONTRUCTOR		AM	PM			
☐ EMPLOYEE ☐ CONTRUCTOR		AM	PM			
☐ EMPLOYEE ☐ CONTRUCTOR		AM	PM			
☐ EMPLOYEE ☐ CONTRUCTOR		AM	PM			
☐ EMPLOYEE ☐ CONTRUCTOR		AM	PM			

EQUIPMENT ON SITE	NO. OF UNITE	WORKING YES / NO

HIRED EQUIPMENT	NO. OF UNITE	EQUIPMENT RENTED	FROM	RATE

NAME: _______________________   SIGNATURE: _______________________

MO TU WE TH FR SA SU
☐   ☐   ☐   ☐   ☐   ☐   ☐

DATE: ____ / ____ / ____

PROJECT:

FOREMAN:

WEATHER

F° ____   C° ____   ____ AM   ____ PM

HOURS DUE TO BAD WEATHER	ISSUED AND DELAYS

NOTE: ____

COMPLETION DATE	DAYS AHEAD OF SCHEDULE	DAYS BEHIND SCHEDULE

SAFETY AND INCIDENTS

SAFETY ISSUES THAT NEED TO BE ADDRESSED	ACCIDENTS / INCIDENTS / STEPS NEEDED TO RESOLVE

SUMMARY OF THE WORK DONE TODAY

IMPORTANT NOTES

NAME	SIGNATURE

TODAY LABOR

INITIALS	TRADE	START	FINISH	PAID HOURS	OVERTIME	COMPANY
☐ EMPLOYEE ☐ CONTRUCTOR		AM	PM			
☐ EMPLOYEE ☐ CONTRUCTOR		AM	PM			
☐ EMPLOYEE ☐ CONTRUCTOR		AM	PM			
☐ EMPLOYEE ☐ CONTRUCTOR		AM	PM			
☐ EMPLOYEE ☐ CONTRUCTOR		AM	PM			
☐ EMPLOYEE ☐ CONTRUCTOR		AM	PM			
☐ EMPLOYEE ☐ CONTRUCTOR		AM	PM			
☐ EMPLOYEE ☐ CONTRUCTOR		AM	PM			

EQUIPMENT ON SITE	NO. OF UNITE	WORKING YES / NO

HIRED EQUIPMENT	NO. OF UNITE	EQUIPMENT RENTED	FROM	RATE

NAME: ___________________________   SIGNATURE: ___________________________

MO TU WE TH FR SA SU
☐ ☐ ☐ ☐ ☐ ☐ ☐

DATE: _____ / _____ / _____

PROJECT:

FOREMAN:

WEATHER

F° _____ C° _____ _____ AM _____ PM

HOURS DUE TO BAD WEATHER

ISSUED AND DELAYS

NOTE: ____

COMPLETION DATE	DAYS AHEAD OF SCHEDULE	DAYS BEHIND SCHEDULE

SAFETY AND INCIDENTS

SAFETY ISSUES THAT NEED TO BE ADDRESSED	ACCIDENTS / INCIDENTS / STEPS NEEDED TO RESOLVE

SUMMARY OF THE WORK DONE TODAY

IMPORTANT NOTES

NAME	SIGNATURE

TODAY LABOR

INITIALS	TRADE	START	FINISH	PAID HOURS	OVERTIME	COMPANY
☐ EMPLOYEE ☐ CONTRUCTOR		AM	PM			
☐ EMPLOYEE ☐ CONTRUCTOR		AM	PM			
☐ EMPLOYEE ☐ CONTRUCTOR		AM	PM			
☐ EMPLOYEE ☐ CONTRUCTOR		AM	PM			
☐ EMPLOYEE ☐ CONTRUCTOR		AM	PM			
☐ EMPLOYEE ☐ CONTRUCTOR		AM	PM			
☐ EMPLOYEE ☐ CONTRUCTOR		AM	PM			
☐ EMPLOYEE ☐ CONTRUCTOR		AM	PM			

EQUIPMENT ON SITE	NO. OF UNITE	WORKING YES / NO

HIRED EQUIPMENT	NO. OF UNITE	EQUIPMENT RENTED	FROM	RATE

NAME: _______________________________   SIGNATURE: _______________________________

MO TU WE TH FR SA SU
☐   ☐   ☐   ☐   ☐   ☐   ☐

DATE: ___ / ___ / ___

PROJECT:

FOREMAN:

WEATHER

F° _____ C° _____ _____ AM _____ PM

HOURS DUE TO BAD WEATHER	ISSUED AND DELAYS

NOTE: ____

COMPLETION DATE	DAYS AHEAD OF SCHEDULE	DAYS BEHIND SCHEDULE

SAFETY AND INCIDENTS

SAFETY ISSUES THAT NEED TO BE ADDRESSED	ACCIDENTS / INCIDENTS / STEPS NEEDED TO RESOLVE

SUMMARY OF THE WORK DONE TODAY

IMPORTANT NOTES

NAME	SIGNATURE

TODAY LABOR

INITIALS	TRADE	START	FINISH	PAID HOURS	OVERTIME	COMPANY
☐ EMPLOYEE ☐ CONTRUCTOR		AM	PM			
☐ EMPLOYEE ☐ CONTRUCTOR		AM	PM			
☐ EMPLOYEE ☐ CONTRUCTOR		AM	PM			
☐ EMPLOYEE ☐ CONTRUCTOR		AM	PM			
☐ EMPLOYEE ☐ CONTRUCTOR		AM	PM			
☐ EMPLOYEE ☐ CONTRUCTOR		AM	PM			
☐ EMPLOYEE ☐ CONTRUCTOR		AM	PM			
☐ EMPLOYEE ☐ CONTRUCTOR		AM	PM			

EQUIPMENT ON SITE	NO. OF UNITE	WORKING YES / NO

HIRED EQUIPMENT	NO. OF UNITE	EQUIPMENT RENTED	FROM	RATE

NAME: _________________________ SIGNATURE: _________________________

MO TU WE TH FR SA SU
☐   ☐   ☐   ☐   ☐   ☐   ☐

DATE: /   /

PROJECT:

FOREMAN:

WEATHER

F°_____ C°_____   _____ AM   _____ PM

HOURS DUE TO BAD WEATHER

ISSUED AND DELAYS

NOTE: ________________________________

COMPLETION DATE	DAYS AHEAD OF SCHEDULE	DAYS BEHIND SCHEDULE

SAFETY AND INCIDENTS

SAFETY ISSUES THAT NEED TO BE ADDRESSED	ACCIDENTS / INCIDENTS / STEPS NEEDED TO RESOLVE

SUMMARY OF THE WORK DONE TODAY

IMPORTANT NOTES

NAME	SIGNATURE

TODAY LABOR

INITIALS	TRADE	START	FINISH	PAID HOURS	OVERTIME	COMPANY
☐ EMPLOYEE ☐ CONTRUCTOR		AM	PM			
☐ EMPLOYEE ☐ CONTRUCTOR		AM	PM			
☐ EMPLOYEE ☐ CONTRUCTOR		AM	PM			
☐ EMPLOYEE ☐ CONTRUCTOR		AM	PM			
☐ EMPLOYEE ☐ CONTRUCTOR		AM	PM			
☐ EMPLOYEE ☐ CONTRUCTOR		AM	PM			
☐ EMPLOYEE ☐ CONTRUCTOR		AM	PM			
☐ EMPLOYEE ☐ CONTRUCTOR		AM	PM			

EQUIPMENT ON SITE	NO. OF UNITE	WORKING YES / NO

HIRED EQUIPMENT	NO. OF UNITE	EQUIPMENT RENTED	FROM	RATE

NAME: _______________________________   SIGNATURE: _______________________________

MO TU WE TH FR SA SU
☐   ☐   ☐   ☐   ☐   ☐   ☐

DATE: ___ / ___ / ___

PROJECT:

FOREMAN:

WEATHER

F°____ C°____ ____AM ____PM

HOURS DUE TO BAD WEATHER	ISSUED AND DELAYS

NOTE: ____

COMPLETION DATE	DAYS AHEAD OF SCHEDULE	DAYS BEHIND SCHEDULE

SAFETY AND INCIDENTS

SAFETY ISSUES THAT NEED TO BE ADDRESSED	ACCIDENTS / INCIDENTS / STEPS NEEDED TO RESOLVE

SUMMARY OF THE WORK DONE TODAY

____
____
____
____

IMPORTANT NOTES

____
____
____
____
____

NAME	SIGNATURE

TODAY LABOR

INITIALS	TRADE	START	FINISH	PAID HOURS	OVERTIME	COMPANY
☐ EMPLOYEE ☐ CONTRUCTOR		AM	PM			
☐ EMPLOYEE ☐ CONTRUCTOR		AM	PM			
☐ EMPLOYEE ☐ CONTRUCTOR		AM	PM			
☐ EMPLOYEE ☐ CONTRUCTOR		AM	PM			
☐ EMPLOYEE ☐ CONTRUCTOR		AM	PM			
☐ EMPLOYEE ☐ CONTRUCTOR		AM	PM			
☐ EMPLOYEE ☐ CONTRUCTOR		AM	PM			
☐ EMPLOYEE ☐ CONTRUCTOR		AM	PM			

EQUIPMENT ON SITE	NO. OF UNITE	WORKING YES / NO

HIRED EQUIPMENT	NO. OF UNITE	EQUIPMENT RENTED	FROM	RATE

NAME: _______________________   SIGNATURE: _______________________

MO TU WE TH FR SA SU
☐   ☐   ☐   ☐   ☐   ☐   ☐

DATE: /   /

PROJECT:

FOREMAN:

WEATHER

F°_____ C°_____   _____ AM _____ PM

HOURS DUE TO BAD WEATHER

ISSUED AND DELAYS

NOTE: ____

COMPLETION DATE	DAYS AHEAD OF SCHEDULE	DAYS BEHIND SCHEDULE

SAFETY AND INCIDENTS

SAFETY ISSUES THAT NEED TO BE ADDRESSED	ACCIDENTS / INCIDENTS / STEPS NEEDED TO RESOLVE

SUMMARY OF THE WORK DONE TODAY

IMPORTANT NOTES

NAME	SIGNATURE

TODAY LABOR

INITIALS	TRADE	START	FINISH	PAID HOURS	OVERTIME	COMPANY
☐ EMPLOYEE ☐ CONTRUCTOR		AM	PM			
☐ EMPLOYEE ☐ CONTRUCTOR		AM	PM			
☐ EMPLOYEE ☐ CONTRUCTOR		AM	PM			
☐ EMPLOYEE ☐ CONTRUCTOR		AM	PM			
☐ EMPLOYEE ☐ CONTRUCTOR		AM	PM			
☐ EMPLOYEE ☐ CONTRUCTOR		AM	PM			
☐ EMPLOYEE ☐ CONTRUCTOR		AM	PM			
☐ EMPLOYEE ☐ CONTRUCTOR		AM	PM			

EQUIPMENT ON SITE	NO. OF UNITE	WORKING YES / NO

HIRED EQUIPMENT	NO. OF UNITE	EQUIPMENT RENTED	FROM	RATE

NAME: _______________________________   SIGNATURE: _______________________________

MO TU WE TH FR SA SU
☐   ☐   ☐   ☐   ☐   ☐   ☐

DATE: /   /

PROJECT:

FOREMAN:

WEATHER

F°______ C°______   ______ AM ______ PM

HOURS DUE TO BAD WEATHER	ISSUED AND DELAYS

NOTE: ______________________________

COMPLETION DATE	DAYS AHEAD OF SCHEDULE	DAYS BEHIND SCHEDULE

SAFETY AND INCIDENTS

SAFETY ISSUES THAT NEED TO BE ADDRESSED	ACCIDENTS / INCIDENTS / STEPS NEEDED TO RESOLVE

SUMMARY OF THE WORK DONE TODAY

IMPORTANT NOTES

NAME	SIGNATURE

TODAY LABOR

INITIALS	TRADE	START	FINISH	PAID HOURS	OVERTIME	COMPANY
☐ EMPLOYEE ☐ CONTRUCTOR		AM	PM			
☐ EMPLOYEE ☐ CONTRUCTOR		AM	PM			
☐ EMPLOYEE ☐ CONTRUCTOR		AM	PM			
☐ EMPLOYEE ☐ CONTRUCTOR		AM	PM			
☐ EMPLOYEE ☐ CONTRUCTOR		AM	PM			
☐ EMPLOYEE ☐ CONTRUCTOR		AM	PM			
☐ EMPLOYEE ☐ CONTRUCTOR		AM	PM			
☐ EMPLOYEE ☐ CONTRUCTOR		AM	PM			

EQUIPMENT ON SITE	NO. OF UNITE	WORKING YES / NO

HIRED EQUIPMENT	NO. OF UNITE	EQUIPMENT RENTED	FROM	RATE

NAME: _______________________   SIGNATURE: _______________________

MO	TU	WE	TH	FR	SA	SU		DATE: / /
☐	☐	☐	☐	☐	☐	☐		

PROJECT:

FOREMAN:

WEATHER F°_____ C°_____ _____AM _____PM

| HOURS DUE TO BAD WEATHER | ISSUED AND DELAYS |

NOTE: _______________

COMPLETION DATE	DAYS AHEAD OF SCHEDULE	DAYS BEHIND SCHEDULE

SAFETY AND INCIDENTS

SAFETY ISSUES THAT NEED TO BE ADDRESSED	ACCIDENTS / INCIDENTS / STEPS NEEDED TO RESOLVE

SUMMARY OF THE WORK DONE TODAY

IMPORTANT NOTES

NAME	SIGNATURE

TODAY LABOR

INITIALS	TRADE	START	FINISH	PAID HOURS	OVERTIME	COMPANY
☐ EMPLOYEE ☐ CONTRUCTOR		AM	PM			
☐ EMPLOYEE ☐ CONTRUCTOR		AM	PM			
☐ EMPLOYEE ☐ CONTRUCTOR		AM	PM			
☐ EMPLOYEE ☐ CONTRUCTOR		AM	PM			
☐ EMPLOYEE ☐ CONTRUCTOR		AM	PM			
☐ EMPLOYEE ☐ CONTRUCTOR		AM	PM			
☐ EMPLOYEE ☐ CONTRUCTOR		AM	PM			
☐ EMPLOYEE ☐ CONTRUCTOR		AM	PM			

EQUIPMENT ON SITE	NO. OF UNITE	WORKING YES / NO

HIRED EQUIPMENT	NO. OF UNITE	EQUIPMENT RENTED	FROM	RATE

NAME: _______________________________   SIGNATURE: _______________________________

MO TU WE TH FR SA SU
☐   ☐   ☐   ☐   ☐   ☐   ☐

DATE: /   /

PROJECT:

FOREMAN:

WEATHER   F°____ C°____ ____AM ____PM

HOURS DUE TO BAD WEATHER

ISSUED AND DELAYS

NOTE: ____

COMPLETION DATE	DAYS AHEAD OF SCHEDULE	DAYS BEHIND SCHEDULE

SAFETY AND INCIDENTS

SAFETY ISSUES THAT NEED TO BE ADDRESSED	ACCIDENTS / INCIDENTS / STEPS NEEDED TO RESOLVE

SUMMARY OF THE WORK DONE TODAY

IMPORTANT NOTES

NAME	SIGNATURE

<table>
<tr><th colspan="8" align="center">TODAY LABOR</th></tr>
<tr><th>INITIALS</th><th>TRADE</th><th>START</th><th>FINISH</th><th>PAID HOURS</th><th>OVERTIME</th><th>COMPANY</th></tr>
<tr><td>☐ EMPLOYEE
☐ CONTRUCTOR</td><td></td><td>AM</td><td>PM</td><td></td><td></td><td></td></tr>
<tr><td>☐ EMPLOYEE
☐ CONTRUCTOR</td><td></td><td>AM</td><td>PM</td><td></td><td></td><td></td></tr>
<tr><td>☐ EMPLOYEE
☐ CONTRUCTOR</td><td></td><td>AM</td><td>PM</td><td></td><td></td><td></td></tr>
<tr><td>☐ EMPLOYEE
☐ CONTRUCTOR</td><td></td><td>AM</td><td>PM</td><td></td><td></td><td></td></tr>
<tr><td>☐ EMPLOYEE
☐ CONTRUCTOR</td><td></td><td>AM</td><td>PM</td><td></td><td></td><td></td></tr>
<tr><td>☐ EMPLOYEE
☐ CONTRUCTOR</td><td></td><td>AM</td><td>PM</td><td></td><td></td><td></td></tr>
<tr><td>☐ EMPLOYEE
☐ CONTRUCTOR</td><td></td><td>AM</td><td>PM</td><td></td><td></td><td></td></tr>
<tr><td>☐ EMPLOYEE
☐ CONTRUCTOR</td><td></td><td>AM</td><td>PM</td><td></td><td></td><td></td></tr>
</table>

EQUIPMENT ON SITE	NO. OF UNITE	WORKING YES / NO

HIRED EQUIPMENT	NO. OF UNITE	EQUIPMENT RENTED	FROM	RATE

NAME: _________________________   SIGNATURE: _________________________

MO	TU	WE	TH	FR	SA	SU
☐	☐	☐	☐	☐	☐	☐

DATE: ___ / ___ / ___

PROJECT:

FOREMAN:

WEATHER F° _____ C° _____ _____ AM _____ PM

HOURS DUE TO BAD WEATHER

ISSUED AND DELAYS

NOTE: _______________________

COMPLETION DATE	DAYS AHEAD OF SCHEDULE	DAYS BEHIND SCHEDULE

SAFETY AND INCIDENTS

SAFETY ISSUES THAT NEED TO BE ADDRESSED	ACCIDENTS / INCIDENTS / STEPS NEEDED TO RESOLVE

SUMMARY OF THE WORK DONE TODAY

IMPORTANT NOTES

NAME	SIGNATURE

TODAY LABOR

INITIALS	TRADE	START	FINISH	PAID HOURS	OVERTIME	COMPANY
☐ EMPLOYEE ☐ CONTRUCTOR		AM	PM			
☐ EMPLOYEE ☐ CONTRUCTOR		AM	PM			
☐ EMPLOYEE ☐ CONTRUCTOR		AM	PM			
☐ EMPLOYEE ☐ CONTRUCTOR		AM	PM			
☐ EMPLOYEE ☐ CONTRUCTOR		AM	PM			
☐ EMPLOYEE ☐ CONTRUCTOR		AM	PM			
☐ EMPLOYEE ☐ CONTRUCTOR		AM	PM			
☐ EMPLOYEE ☐ CONTRUCTOR		AM	PM			

EQUIPMENT ON SITE	NO. OF UNITE	WORKING YES / NO

HIRED EQUIPMENT	NO. OF UNITE	EQUIPMENT RENTED	FROM	RATE

NAME: _______________________ SIGNATURE: _______________________

MO TU WE TH FR SA SU
☐   ☐   ☐   ☐   ☐   ☐   ☐

DATE: ____ / ____ / ____

PROJECT:

FOREMAN:

WEATHER

F°_____ C°_____   _____ AM   _____ PM

HOURS DUE TO BAD WEATHER	ISSUED AND DELAYS

NOTE: ____________________________________

COMPLETION DATE	DAYS AHEAD OF SCHEDULE	DAYS BEHIND SCHEDULE

SAFETY AND INCIDENTS

SAFETY ISSUES THAT NEED TO BE ADDRESSED	ACCIDENTS / INCIDENTS / STEPS NEEDED TO RESOLVE

SUMMARY OF THE WORK DONE TODAY

IMPORTANT NOTES

NAME	SIGNATURE

TODAY LABOR

INITIALS	TRADE	START	FINISH	PAID HOURS	OVERTIME	COMPANY
☐ EMPLOYEE ☐ CONTRUCTOR		AM	PM			
☐ EMPLOYEE ☐ CONTRUCTOR		AM	PM			
☐ EMPLOYEE ☐ CONTRUCTOR		AM	PM			
☐ EMPLOYEE ☐ CONTRUCTOR		AM	PM			
☐ EMPLOYEE ☐ CONTRUCTOR		AM	PM			
☐ EMPLOYEE ☐ CONTRUCTOR		AM	PM			
☐ EMPLOYEE ☐ CONTRUCTOR		AM	PM			
☐ EMPLOYEE ☐ CONTRUCTOR		AM	PM			

EQUIPMENT ON SITE	NO. OF UNITE	WORKING YES / NO

HIRED EQUIPMENT	NO. OF UNITE	EQUIPMENT RENTED	FROM	RATE

NAME: _______________________   SIGNATURE: _______________________

MO   TU   WE   TH   FR   SA   SU
☐   ☐   ☐   ☐   ☐   ☐   ☐

DATE: /   /

PROJECT:

FOREMAN:

WEATHER

F°____   C°____   ____ AM   ____ PM

HOURS DUE TO BAD WEATHER	ISSUED AND DELAYS

NOTE: ____

COMPLETION DATE	DAYS AHEAD OF SCHEDULE	DAYS BEHIND SCHEDULE

SAFETY AND INCIDENTS

SAFETY ISSUES THAT NEED TO BE ADDRESSED	ACCIDENTS / INCIDENTS / STEPS NEEDED TO RESOLVE

SUMMARY OF THE WORK DONE TODAY

IMPORTANT NOTES

NAME	SIGNATURE

TODAY LABOR

INITIALS	TRADE	START	FINISH	PAID HOURS	OVERTIME	COMPANY
☐ EMPLOYEE ☐ CONTRUCTOR		AM	PM			
☐ EMPLOYEE ☐ CONTRUCTOR		AM	PM			
☐ EMPLOYEE ☐ CONTRUCTOR		AM	PM			
☐ EMPLOYEE ☐ CONTRUCTOR		AM	PM			
☐ EMPLOYEE ☐ CONTRUCTOR		AM	PM			
☐ EMPLOYEE ☐ CONTRUCTOR		AM	PM			
☐ EMPLOYEE ☐ CONTRUCTOR		AM	PM			
☐ EMPLOYEE ☐ CONTRUCTOR		AM	PM			

EQUIPMENT ON SITE	NO. OF UNITE	WORKING YES / NO

HIRED EQUIPMENT	NO. OF UNITE	EQUIPMENT RENTED	FROM	RATE

NAME: _______________________   SIGNATURE: _______________________

MO TU WE TH FR SA SU
☐   ☐   ☐   ☐   ☐   ☐   ☐

DATE: / /

PROJECT:

FOREMAN:

WEATHER

F°____ C°____   ____AM ____PM

| HOURS DUE TO BAD WEATHER | ISSUED AND DELAYS |

NOTE: ____________________

| COMPLETION DATE | DAYS AHEAD OF SCHEDULE | DAYS BEHIND SCHEDULE |

SAFETY AND INCIDENTS

| SAFETY ISSUES THAT NEED TO BE ADDRESSED | ACCIDENTS / INCIDENTS / STEPS NEEDED TO RESOLVE |

SUMMARY OF THE WORK DONE TODAY

IMPORTANT NOTES

| NAME | SIGNATURE |

TODAY LABOR

INITIALS	TRADE	START	FINISH	PAID HOURS	OVERTIME	COMPANY
☐ EMPLOYEE ☐ CONTRUCTOR		AM	PM			
☐ EMPLOYEE ☐ CONTRUCTOR		AM	PM			
☐ EMPLOYEE ☐ CONTRUCTOR		AM	PM			
☐ EMPLOYEE ☐ CONTRUCTOR		AM	PM			
☐ EMPLOYEE ☐ CONTRUCTOR		AM	PM			
☐ EMPLOYEE ☐ CONTRUCTOR		AM	PM			
☐ EMPLOYEE ☐ CONTRUCTOR		AM	PM			
☐ EMPLOYEE ☐ CONTRUCTOR		AM	PM			

EQUIPMENT ON SITE	NO. OF UNITE	WORKING YES / NO

HIRED EQUIPMENT	NO. OF UNITE	EQUIPMENT RENTED	FROM	RATE

NAME: ___________________________   SIGNATURE: ___________________________

MO TU WE TH FR SA SU
☐   ☐   ☐   ☐   ☐   ☐   ☐

DATE: /   /

PROJECT:

FOREMAN:

WEATHER

F°_____ C°_____   _____ AM   _____ PM

HOURS DUE TO BAD WEATHER

ISSUED AND DELAYS

NOTE: _______________________________________

COMPLETION DATE	DAYS AHEAD OF SCHEDULE	DAYS BEHIND SCHEDULE

SAFETY AND INCIDENTS

SAFETY ISSUES THAT NEED TO BE ADDRESSED	ACCIDENTS / INCIDENTS / STEPS NEEDED TO RESOLVE

SUMMARY OF THE WORK DONE TODAY

IMPORTANT NOTES

NAME	SIGNATURE

TODAY LABOR

INITIALS	TRADE	START	FINISH	PAID HOURS	OVERTIME	COMPANY
☐ EMPLOYEE ☐ CONTRUCTOR		AM	PM			
☐ EMPLOYEE ☐ CONTRUCTOR		AM	PM			
☐ EMPLOYEE ☐ CONTRUCTOR		AM	PM			
☐ EMPLOYEE ☐ CONTRUCTOR		AM	PM			
☐ EMPLOYEE ☐ CONTRUCTOR		AM	PM			
☐ EMPLOYEE ☐ CONTRUCTOR		AM	PM			
☐ EMPLOYEE ☐ CONTRUCTOR		AM	PM			
☐ EMPLOYEE ☐ CONTRUCTOR		AM	PM			

EQUIPMENT ON SITE	NO. OF UNITE	WORKING YES / NO

HIRED EQUIPMENT	NO. OF UNITE	EQUIPMENT RENTED	FROM	RATE

NAME: _______________________   SIGNATURE: _______________________

MO   TU   WE   TH   FR   SA   SU
☐   ☐   ☐   ☐   ☐   ☐   ☐

DATE: /   /

PROJECT:

FOREMAN:

WEATHER   F° _____   C° _____   _____ AM   _____ PM

HOURS DUE TO BAD WEATHER

ISSUED AND DELAYS

NOTE: ____

COMPLETION DATE	DAYS AHEAD OF SCHEDULE	DAYS BEHIND SCHEDULE

SAFETY AND INCIDENTS

SAFETY ISSUES THAT NEED TO BE ADDRESSED	ACCIDENTS / INCIDENTS / STEPS NEEDED TO RESOLVE

SUMMARY OF THE WORK DONE TODAY

IMPORTANT NOTES

NAME	SIGNATURE

TODAY LABOR

INITIALS	TRADE	START	FINISH	PAID HOURS	OVERTIME	COMPANY
☐ EMPLOYEE ☐ CONTRUCTOR		AM	PM			
☐ EMPLOYEE ☐ CONTRUCTOR		AM	PM			
☐ EMPLOYEE ☐ CONTRUCTOR		AM	PM			
☐ EMPLOYEE ☐ CONTRUCTOR		AM	PM			
☐ EMPLOYEE ☐ CONTRUCTOR		AM	PM			
☐ EMPLOYEE ☐ CONTRUCTOR		AM	PM			
☐ EMPLOYEE ☐ CONTRUCTOR		AM	PM			
☐ EMPLOYEE ☐ CONTRUCTOR		AM	PM			

EQUIPMENT ON SITE	NO. OF UNITE	WORKING YES / NO

HIRED EQUIPMENT	NO. OF UNITE	EQUIPMENT RENTED	FROM	RATE

NAME: ________________________   SIGNATURE: ________________________

MO TU WE TH FR SA SU
☐   ☐   ☐   ☐   ☐   ☐   ☐

DATE: ___/___/___

PROJECT:

FOREMAN:

WEATHER   F°______ C°______   ______ AM   ______ PM

HOURS DUE TO BAD WEATHER	ISSUED AND DELAYS

NOTE: ____

COMPLETION DATE	DAYS AHEAD OF SCHEDULE	DAYS BEHIND SCHEDULE

SAFETY AND INCIDENTS

SAFETY ISSUES THAT NEED TO BE ADDRESSED	ACCIDENTS / INCIDENTS / STEPS NEEDED TO RESOLVE

SUMMARY OF THE WORK DONE TODAY

IMPORTANT NOTES

NAME	SIGNATURE

INITIALS	TRADE	START	FINISH	PAID HOURS	OVERTIME	COMPANY
☐ EMPLOYEE ☐ CONTRUCTOR		AM	PM			
☐ EMPLOYEE ☐ CONTRUCTOR		AM	PM			
☐ EMPLOYEE ☐ CONTRUCTOR		AM	PM			
☐ EMPLOYEE ☐ CONTRUCTOR		AM	PM			
☐ EMPLOYEE ☐ CONTRUCTOR		AM	PM			
☐ EMPLOYEE ☐ CONTRUCTOR		AM	PM			
☐ EMPLOYEE ☐ CONTRUCTOR		AM	PM			
☐ EMPLOYEE ☐ CONTRUCTOR		AM	PM			

EQUIPMENT ON SITE	NO. OF UNITE	WORKING YES / NO

HIRED EQUIPMENT	NO. OF UNITE	EQUIPMENT RENTED	FROM	RATE

NAME: _______________________________

SIGNATURE: _______________________________

MO TU WE TH FR SA SU
☐   ☐   ☐   ☐   ☐   ☐   ☐

DATE: /   /

PROJECT:

FOREMAN:

WEATHER

F°_____ C°_____   _____ AM _____ PM

HOURS DUE TO BAD WEATHER

ISSUED AND DELAYS

NOTE: _______________________________

COMPLETION DATE	DAYS AHEAD OF SCHEDULE	DAYS BEHIND SCHEDULE

SAFETY AND INCIDENTS

SAFETY ISSUES THAT NEED TO BE ADDRESSED	ACCIDENTS / INCIDENTS / STEPS NEEDED TO RESOLVE

SUMMARY OF THE WORK DONE TODAY

IMPORTANT NOTES

NAME	SIGNATURE

TODAY LABOR

INITIALS	TRADE	START	FINISH	PAID HOURS	OVERTIME	COMPANY
☐ EMPLOYEE ☐ CONTRUCTOR		AM	PM			
☐ EMPLOYEE ☐ CONTRUCTOR		AM	PM			
☐ EMPLOYEE ☐ CONTRUCTOR		AM	PM			
☐ EMPLOYEE ☐ CONTRUCTOR		AM	PM			
☐ EMPLOYEE ☐ CONTRUCTOR		AM	PM			
☐ EMPLOYEE ☐ CONTRUCTOR		AM	PM			
☐ EMPLOYEE ☐ CONTRUCTOR		AM	PM			
☐ EMPLOYEE ☐ CONTRUCTOR		AM	PM			

EQUIPMENT ON SITE	NO. OF UNITE	WORKING YES / NO

HIRED EQUIPMENT	NO. OF UNITE	EQUIPMENT RENTED	FROM	RATE

NAME: _______________________________   SIGNATURE: _______________________________

MO TU WE TH FR SA SU
☐   ☐   ☐   ☐   ☐   ☐   ☐

DATE: ___/___/___

PROJECT:

FOREMAN:

WEATHER

F°_____ C°_____   _____ AM _____ PM

HOURS DUE TO BAD WEATHER

ISSUED AND DELAYS

NOTE: _______________________________

COMPLETION DATE	DAYS AHEAD OF SCHEDULE	DAYS BEHIND SCHEDULE

SAFETY AND INCIDENTS

SAFETY ISSUES THAT NEED TO BE ADDRESSED	ACCIDENTS / INCIDENTS / STEPS NEEDED TO RESOLVE

SUMMARY OF THE WORK DONE TODAY

IMPORTANT NOTES

NAME	SIGNATURE

TODAY LABOR

INITIALS	TRADE	START	FINISH	PAID HOURS	OVERTIME	COMPANY
☐ EMPLOYEE ☐ CONTRUCTOR		AM	PM			
☐ EMPLOYEE ☐ CONTRUCTOR		AM	PM			
☐ EMPLOYEE ☐ CONTRUCTOR		AM	PM			
☐ EMPLOYEE ☐ CONTRUCTOR		AM	PM			
☐ EMPLOYEE ☐ CONTRUCTOR		AM	PM			
☐ EMPLOYEE ☐ CONTRUCTOR		AM	PM			
☐ EMPLOYEE ☐ CONTRUCTOR		AM	PM			
☐ EMPLOYEE ☐ CONTRUCTOR		AM	PM			

EQUIPMENT ON SITE	NO. OF UNITE	WORKING YES / NO

HIRED EQUIPMENT	NO. OF UNITE	EQUIPMENT RENTED	FROM	RATE

NAME: _______________________   SIGNATURE: _______________________

MO TU WE TH FR SA SU
☐   ☐   ☐   ☐   ☐   ☐   ☐

DATE: /   /

PROJECT:

FOREMAN:

WEATHER

F°______ C°______   ______ AM   ______ PM

HOURS DUE TO BAD WEATHER

ISSUED AND DELAYS

NOTE: ____

COMPLETION DATE	DAYS AHEAD OF SCHEDULE	DAYS BEHIND SCHEDULE

SAFETY AND INCIDENTS

SAFETY ISSUES THAT NEED TO BE ADDRESSED	ACCIDENTS / INCIDENTS / STEPS NEEDED TO RESOLVE

SUMMARY OF THE WORK DONE TODAY

IMPORTANT NOTES

NAME	SIGNATURE

TODAY LABOR

INITIALS	TRADE	START	FINISH	PAID HOURS	OVERTIME	COMPANY
☐ EMPLOYEE ☐ CONTRUCTOR		AM	PM			
☐ EMPLOYEE ☐ CONTRUCTOR		AM	PM			
☐ EMPLOYEE ☐ CONTRUCTOR		AM	PM			
☐ EMPLOYEE ☐ CONTRUCTOR		AM	PM			
☐ EMPLOYEE ☐ CONTRUCTOR		AM	PM			
☐ EMPLOYEE ☐ CONTRUCTOR		AM	PM			
☐ EMPLOYEE ☐ CONTRUCTOR		AM	PM			
☐ EMPLOYEE ☐ CONTRUCTOR		AM	PM			

EQUIPMENT ON SITE	NO. OF UNITE	WORKING YES / NO

HIRED EQUIPMENT	NO. OF UNITE	EQUIPMENT RENTED	FROM	RATE

NAME: _______________________   SIGNATURE: _______________________

MO TU WE TH FR SA SU
☐   ☐   ☐   ☐   ☐   ☐   ☐

DATE: ___/___/___

PROJECT:

FOREMAN:

WEATHER

F°____ C°____   ____ AM ____ PM

HOURS DUE TO BAD WEATHER

ISSUED AND DELAYS

NOTE: _______________________________________

COMPLETION DATE	DAYS AHEAD OF SCHEDULE	DAYS BEHIND SCHEDULE

SAFETY AND INCIDENTS

SAFETY ISSUES THAT NEED TO BE ADDRESSED	ACCIDENTS / INCIDENTS / STEPS NEEDED TO RESOLVE

SUMMARY OF THE WORK DONE TODAY

IMPORTANT NOTES

NAME	SIGNATURE

TODAY LABOR

INITIALS	TRADE	START	FINISH	PAID HOURS	OVERTIME	COMPANY
☐ EMPLOYEE ☐ CONTRUCTOR		AM	PM			
☐ EMPLOYEE ☐ CONTRUCTOR		AM	PM			
☐ EMPLOYEE ☐ CONTRUCTOR		AM	PM			
☐ EMPLOYEE ☐ CONTRUCTOR		AM	PM			
☐ EMPLOYEE ☐ CONTRUCTOR		AM	PM			
☐ EMPLOYEE ☐ CONTRUCTOR		AM	PM			
☐ EMPLOYEE ☐ CONTRUCTOR		AM	PM			
☐ EMPLOYEE ☐ CONTRUCTOR		AM	PM			

EQUIPMENT ON SITE	NO. OF UNITE	WORKING YES / NO

HIRED EQUIPMENT	NO. OF UNITE	EQUIPMENT RENTED	FROM	RATE

NAME: _______________________   SIGNATURE: _______________________

MO TU WE TH FR SA SU
☐   ☐   ☐   ☐   ☐   ☐   ☐

DATE: ___/___/___

PROJECT:

FOREMAN:

WEATHER   F°___   C°___   ___ AM   ___ PM

HOURS DUE TO BAD WEATHER	ISSUED AND DELAYS

NOTE: ___________________________

COMPLETION DATE	DAYS AHEAD OF SCHEDULE	DAYS BEHIND SCHEDULE

SAFETY AND INCIDENTS

SAFETY ISSUES THAT NEED TO BE ADDRESSED	ACCIDENTS / INCIDENTS / STEPS NEEDED TO RESOLVE

SUMMARY OF THE WORK DONE TODAY

IMPORTANT NOTES

NAME	SIGNATURE

TODAY LABOR

INITIALS	TRADE	START	FINISH	PAID HOURS	OVERTIME	COMPANY
☐ EMPLOYEE ☐ CONTRUCTOR		AM	PM			
☐ EMPLOYEE ☐ CONTRUCTOR		AM	PM			
☐ EMPLOYEE ☐ CONTRUCTOR		AM	PM			
☐ EMPLOYEE ☐ CONTRUCTOR		AM	PM			
☐ EMPLOYEE ☐ CONTRUCTOR		AM	PM			
☐ EMPLOYEE ☐ CONTRUCTOR		AM	PM			
☐ EMPLOYEE ☐ CONTRUCTOR		AM	PM			
☐ EMPLOYEE ☐ CONTRUCTOR		AM	PM			

EQUIPMENT ON SITE	NO. OF UNITE	WORKING YES / NO

HIRED EQUIPMENT	NO. OF UNITE	EQUIPMENT RENTED	FROM	RATE

NAME: _____________________________   SIGNATURE: _____________________________

MO TU WE TH FR SA SU
☐   ☐   ☐   ☐   ☐   ☐   ☐

DATE: /   /

PROJECT:

FOREMAN:

WEATHER

F°_____ C°_____   _____AM _____PM

HOURS DUE TO BAD WEATHER

ISSUED AND DELAYS

NOTE: ____

COMPLETION DATE	DAYS AHEAD OF SCHEDULE	DAYS BEHIND SCHEDULE

SAFETY AND INCIDENTS

SAFETY ISSUES THAT NEED TO BE ADDRESSED	ACCIDENTS / INCIDENTS / STEPS NEEDED TO RESOLVE

SUMMARY OF THE WORK DONE TODAY

IMPORTANT NOTES

NAME	SIGNATURE

TODAY LABOR

INITIALS	TRADE	START	FINISH	PAID HOURS	OVERTIME	COMPANY
☐ EMPLOYEE ☐ CONTRUCTOR		AM	PM			
☐ EMPLOYEE ☐ CONTRUCTOR		AM	PM			
☐ EMPLOYEE ☐ CONTRUCTOR		AM	PM			
☐ EMPLOYEE ☐ CONTRUCTOR		AM	PM			
☐ EMPLOYEE ☐ CONTRUCTOR		AM	PM			
☐ EMPLOYEE ☐ CONTRUCTOR		AM	PM			
☐ EMPLOYEE ☐ CONTRUCTOR		AM	PM			
☐ EMPLOYEE ☐ CONTRUCTOR		AM	PM			

EQUIPMENT ON SITE	NO. OF UNITE	WORKING YES / NO

HIRED EQUIPMENT	NO. OF UNITE	EQUIPMENT RENTED	FROM	RATE

NAME: _______________________   SIGNATURE: _______________________

MO ☐　TU ☐　WE ☐　TH ☐　FR ☐　SA ☐　SU ☐

DATE: ___ / ___ / ___

PROJECT:

FOREMAN:

WEATHER　F° _____　C° _____　_____ AM　_____ PM

HOURS DUE TO BAD WEATHER

ISSUED AND DELAYS

NOTE: _______________________________________

COMPLETION DATE	DAYS AHEAD OF SCHEDULE	DAYS BEHIND SCHEDULE

SAFETY AND INCIDENTS

SAFETY ISSUES THAT NEED TO BE ADDRESSED	ACCIDENTS / INCIDENTS / STEPS NEEDED TO RESOLVE

SUMMARY OF THE WORK DONE TODAY

IMPORTANT NOTES

NAME	SIGNATURE

TODAY LABOR

INITIALS	TRADE	START	FINISH	PAID HOURS	OVERTIME	COMPANY
☐ EMPLOYEE ☐ CONTRUCTOR		AM	PM			
☐ EMPLOYEE ☐ CONTRUCTOR		AM	PM			
☐ EMPLOYEE ☐ CONTRUCTOR		AM	PM			
☐ EMPLOYEE ☐ CONTRUCTOR		AM	PM			
☐ EMPLOYEE ☐ CONTRUCTOR		AM	PM			
☐ EMPLOYEE ☐ CONTRUCTOR		AM	PM			
☐ EMPLOYEE ☐ CONTRUCTOR		AM	PM			
☐ EMPLOYEE ☐ CONTRUCTOR		AM	PM			

EQUIPMENT ON SITE	NO. OF UNITE	WORKING YES / NO

HIRED EQUIPMENT	NO. OF UNITE	EQUIPMENT RENTED	FROM	RATE

NAME: _______________________   SIGNATURE: _______________________

MO   TU   WE   TH   FR   SA   SU   DATE: /   /

☐   ☐   ☐   ☐   ☐   ☐   ☐

PROJECT: **FOREMAN:**

WEATHER

F°_____ C°_____   _____ AM   _____ PM

HOURS DUE TO BAD WEATHER	ISSUED AND DELAYS

NOTE: ____

COMPLETION DATE	DAYS AHEAD OF SCHEDULE	DAYS BEHIND SCHEDULE

SAFETY AND INCIDENTS

SAFETY ISSUES THAT NEED TO BE ADDRESSED	ACCIDENTS / INCIDENTS / STEPS NEEDED TO RESOLVE

SUMMARY OF THE WORK DONE TODAY

IMPORTANT NOTES

NAME	SIGNATURE

TODAY LABOR

INITIALS	TRADE	START	FINISH	PAID HOURS	OVERTIME	COMPANY
☐ EMPLOYEE ☐ CONTRUCTOR		AM	PM			
☐ EMPLOYEE ☐ CONTRUCTOR		AM	PM			
☐ EMPLOYEE ☐ CONTRUCTOR		AM	PM			
☐ EMPLOYEE ☐ CONTRUCTOR		AM	PM			
☐ EMPLOYEE ☐ CONTRUCTOR		AM	PM			
☐ EMPLOYEE ☐ CONTRUCTOR		AM	PM			
☐ EMPLOYEE ☐ CONTRUCTOR		AM	PM			
☐ EMPLOYEE ☐ CONTRUCTOR		AM	PM			

EQUIPMENT ON SITE	NO. OF UNITE	WORKING YES / NO

HIRED EQUIPMENT	NO. OF UNITE	EQUIPMENT RENTED	FROM	RATE

NAME: _____________________________ SIGNATURE: _____________________________

MO TU WE TH FR SA SU
☐   ☐   ☐   ☐   ☐   ☐   ☐

DATE: /   /

PROJECT:

FOREMAN:

WEATHER

F°____ C°____   ____ AM ____ PM

| HOURS DUE TO BAD WEATHER | ISSUED AND DELAYS |

NOTE: ____

COMPLETION DATE	DAYS AHEAD OF SCHEDULE	DAYS BEHIND SCHEDULE

SAFETY AND INCIDENTS

SAFETY ISSUES THAT NEED TO BE ADDRESSED	ACCIDENTS / INCIDENTS / STEPS NEEDED TO RESOLVE

SUMMARY OF THE WORK DONE TODAY

IMPORTANT NOTES

NAME	SIGNATURE

TODAY LABOR

INITIALS	TRADE	START	FINISH	PAID HOURS	OVERTIME	COMPANY
☐ EMPLOYEE ☐ CONTRUCTOR		AM	PM			
☐ EMPLOYEE ☐ CONTRUCTOR		AM	PM			
☐ EMPLOYEE ☐ CONTRUCTOR		AM	PM			
☐ EMPLOYEE ☐ CONTRUCTOR		AM	PM			
☐ EMPLOYEE ☐ CONTRUCTOR		AM	PM			
☐ EMPLOYEE ☐ CONTRUCTOR		AM	PM			
☐ EMPLOYEE ☐ CONTRUCTOR		AM	PM			
☐ EMPLOYEE ☐ CONTRUCTOR		AM	PM			

EQUIPMENT ON SITE	NO. OF UNITE	WORKING YES / NO

HIRED EQUIPMENT	NO. OF UNITE	EQUIPMENT RENTED	FROM	RATE

NAME: _______________________   SIGNATURE: _______________________

MO	TU	WE	TH	FR	SA	SU
☐	☐	☐	☐	☐	☐	☐

DATE: ___ / ___ / ___

PROJECT:

FOREMAN:

WEATHER

F° _____ C° _____ _____ AM _____ PM

HOURS DUE TO BAD WEATHER

ISSUED AND DELAYS

NOTE: _______________________________

COMPLETION DATE	DAYS AHEAD OF SCHEDULE	DAYS BEHIND SCHEDULE

SAFETY AND INCIDENTS

SAFETY ISSUES THAT NEED TO BE ADDRESSED	ACCIDENTS / INCIDENTS / STEPS NEEDED TO RESOLVE

SUMMARY OF THE WORK DONE TODAY

IMPORTANT NOTES

NAME	SIGNATURE

TODAY LABOR

INITIALS	TRADE	START	FINISH	PAID HOURS	OVERTIME	COMPANY
☐ EMPLOYEE ☐ CONTRUCTOR		AM	PM			
☐ EMPLOYEE ☐ CONTRUCTOR		AM	PM			
☐ EMPLOYEE ☐ CONTRUCTOR		AM	PM			
☐ EMPLOYEE ☐ CONTRUCTOR		AM	PM			
☐ EMPLOYEE ☐ CONTRUCTOR		AM	PM			
☐ EMPLOYEE ☐ CONTRUCTOR		AM	PM			
☐ EMPLOYEE ☐ CONTRUCTOR		AM	PM			
☐ EMPLOYEE ☐ CONTRUCTOR		AM	PM			

EQUIPMENT ON SITE	NO. OF UNITE	WORKING YES / NO

HIRED EQUIPMENT	NO. OF UNITE	EQUIPMENT RENTED	FROM	RATE

NAME: ______________________________   SIGNATURE: ______________________________

MO TU WE TH FR SA SU

DATE: / /

PROJECT:

FOREMAN:

WEATHER

F° ___ C° ___   ___ AM ___ PM

HOURS DUE TO BAD WEATHER

ISSUED AND DELAYS

NOTE: ___

COMPLETION DATE	DAYS AHEAD OF SCHEDULE	DAYS BEHIND SCHEDULE

SAFETY AND INCIDENTS

SAFETY ISSUES THAT NEED TO BE ADDRESSED	ACCIDENTS / INCIDENTS / STEPS NEEDED TO RESOLVE

SUMMARY OF THE WORK DONE TODAY

IMPORTANT NOTES

NAME	SIGNATURE

TODAY LABOR

INITIALS	TRADE	START	FINISH	PAID HOURS	OVERTIME	COMPANY
☐ EMPLOYEE ☐ CONTRUCTOR		AM	PM			
☐ EMPLOYEE ☐ CONTRUCTOR		AM	PM			
☐ EMPLOYEE ☐ CONTRUCTOR		AM	PM			
☐ EMPLOYEE ☐ CONTRUCTOR		AM	PM			
☐ EMPLOYEE ☐ CONTRUCTOR		AM	PM			
☐ EMPLOYEE ☐ CONTRUCTOR		AM	PM			
☐ EMPLOYEE ☐ CONTRUCTOR		AM	PM			
☐ EMPLOYEE ☐ CONTRUCTOR		AM	PM			

EQUIPMENT ON SITE	NO. OF UNITE	WORKING YES / NO

HIRED EQUIPMENT	NO. OF UNITE	EQUIPMENT RENTED	FROM	RATE

NAME: _______________________   SIGNATURE: _______________________

MO	TU	WE	TH	FR	SA	SU		DATE: / /
☐	☐	☐	☐	☐	☐	☐		

PROJECT:

FOREMAN:

WEATHER

F° _____ C° _____ _____ AM _____ PM

| HOURS DUE TO BAD WEATHER | ISSUED AND DELAYS |

NOTE: ____

COMPLETION DATE	DAYS AHEAD OF SCHEDULE	DAYS BEHIND SCHEDULE

SAFETY AND INCIDENTS

SAFETY ISSUES THAT NEED TO BE ADDRESSED	ACCIDENTS / INCIDENTS / STEPS NEEDED TO RESOLVE

SUMMARY OF THE WORK DONE TODAY

IMPORTANT NOTES

NAME	SIGNATURE

TODAY LABOR

INITIALS	TRADE	START	FINISH	PAID HOURS	OVERTIME	COMPANY
☐ EMPLOYEE ☐ CONTRUCTOR		AM	PM			
☐ EMPLOYEE ☐ CONTRUCTOR		AM	PM			
☐ EMPLOYEE ☐ CONTRUCTOR		AM	PM			
☐ EMPLOYEE ☐ CONTRUCTOR		AM	PM			
☐ EMPLOYEE ☐ CONTRUCTOR		AM	PM			
☐ EMPLOYEE ☐ CONTRUCTOR		AM	PM			
☐ EMPLOYEE ☐ CONTRUCTOR		AM	PM			
☐ EMPLOYEE ☐ CONTRUCTOR		AM	PM			

EQUIPMENT ON SITE	NO. OF UNITE	WORKING YES / NO

HIRED EQUIPMENT	NO. OF UNITE	EQUIPMENT RENTED	FROM	RATE

NAME: _______________________   SIGNATURE: _______________________

MO TU WE TH FR SA SU
☐   ☐   ☐   ☐   ☐   ☐   ☐

DATE: /   /

PROJECT:

FOREMAN:

WEATHER   F°_____ C°_____   _____ AM _____ PM

HOURS DUE TO BAD WEATHER

ISSUED AND DELAYS

NOTE: _______________________________________

COMPLETION DATE	DAYS AHEAD OF SCHEDULE	DAYS BEHIND SCHEDULE

SAFETY AND INCIDENTS

SAFETY ISSUES THAT NEED TO BE ADDRESSED	ACCIDENTS / INCIDENTS / STEPS NEEDED TO RESOLVE

SUMMARY OF THE WORK DONE TODAY

IMPORTANT NOTES

NAME	SIGNATURE

TODAY LABOR

INITIALS	TRADE	START	FINISH	PAID HOURS	OVERTIME	COMPANY
☐ EMPLOYEE ☐ CONTRUCTOR		AM	PM			
☐ EMPLOYEE ☐ CONTRUCTOR		AM	PM			
☐ EMPLOYEE ☐ CONTRUCTOR		AM	PM			
☐ EMPLOYEE ☐ CONTRUCTOR		AM	PM			
☐ EMPLOYEE ☐ CONTRUCTOR		AM	PM			
☐ EMPLOYEE ☐ CONTRUCTOR		AM	PM			
☐ EMPLOYEE ☐ CONTRUCTOR		AM	PM			
☐ EMPLOYEE ☐ CONTRUCTOR		AM	PM			

EQUIPMENT ON SITE	NO. OF UNITE	WORKING YES / NO

HIRED EQUIPMENT	NO. OF UNITE	EQUIPMENT RENTED	FROM	RATE

NAME: _________________________   SIGNATURE: _________________________

MO TU WE TH FR SA SU
☐   ☐   ☐   ☐   ☐   ☐   ☐

DATE: ___ / ___ / ___

PROJECT:

FOREMAN:

WEATHER   F° ____ C° ____   ____ AM ____ PM

HOURS DUE TO BAD WEATHER	ISSUED AND DELAYS

NOTE: ____

COMPLETION DATE	DAYS AHEAD OF SCHEDULE	DAYS BEHIND SCHEDULE

SAFETY AND INCIDENTS

SAFETY ISSUES THAT NEED TO BE ADDRESSED	ACCIDENTS / INCIDENTS / STEPS NEEDED TO RESOLVE

SUMMARY OF THE WORK DONE TODAY

____
____
____
____
____

IMPORTANT NOTES

____
____
____
____
____

NAME	SIGNATURE

TODAY LABOR

INITIALS	TRADE	START	FINISH	PAID HOURS	OVERTIME	COMPANY
☐ EMPLOYEE ☐ CONTRUCTOR		AM	PM			
☐ EMPLOYEE ☐ CONTRUCTOR		AM	PM			
☐ EMPLOYEE ☐ CONTRUCTOR		AM	PM			
☐ EMPLOYEE ☐ CONTRUCTOR		AM	PM			
☐ EMPLOYEE ☐ CONTRUCTOR		AM	PM			
☐ EMPLOYEE ☐ CONTRUCTOR		AM	PM			
☐ EMPLOYEE ☐ CONTRUCTOR		AM	PM			
☐ EMPLOYEE ☐ CONTRUCTOR		AM	PM			

EQUIPMENT ON SITE	NO. OF UNITE	WORKING YES / NO

HIRED EQUIPMENT	NO. OF UNITE	EQUIPMENT RENTED	FROM	RATE

NAME: _________________________   SIGNATURE: _________________________

MO TU WE TH FR SA SU
☐   ☐   ☐   ☐   ☐   ☐   ☐

DATE: /   /

PROJECT:

FOREMAN:

WEATHER

F°_____ C°_____   _____ AM _____ PM

HOURS DUE TO BAD WEATHER

ISSUED AND DELAYS

NOTE: _______________________________________

COMPLETION DATE	DAYS AHEAD OF SCHEDULE	DAYS BEHIND SCHEDULE

SAFETY AND INCIDENTS

SAFETY ISSUES THAT NEED TO BE ADDRESSED	ACCIDENTS / INCIDENTS / STEPS NEEDED TO RESOLVE

SUMMARY OF THE WORK DONE TODAY

IMPORTANT NOTES

NAME	SIGNATURE

TODAY LABOR

INITIALS	TRADE	START	FINISH	PAID HOURS	OVERTIME	COMPANY
☐ EMPLOYEE ☐ CONTRUCTOR		AM	PM			
☐ EMPLOYEE ☐ CONTRUCTOR		AM	PM			
☐ EMPLOYEE ☐ CONTRUCTOR		AM	PM			
☐ EMPLOYEE ☐ CONTRUCTOR		AM	PM			
☐ EMPLOYEE ☐ CONTRUCTOR		AM	PM			
☐ EMPLOYEE ☐ CONTRUCTOR		AM	PM			
☐ EMPLOYEE ☐ CONTRUCTOR		AM	PM			
☐ EMPLOYEE ☐ CONTRUCTOR		AM	PM			

EQUIPMENT ON SITE	NO. OF UNITE	WORKING YES / NO

HIRED EQUIPMENT	NO. OF UNITE	EQUIPMENT RENTED	FROM	RATE

NAME: _______________________   SIGNATURE: _______________________

| MO | TU | WE | TH | FR | SA | SU | | DATE: / / |
| ☐ | ☐ | ☐ | ☐ | ☐ | ☐ | ☐ | | |

| PROJECT: | FOREMAN: |

WEATHER F°____ C°____ ____AM ____PM

| HOURS DUE TO BAD WEATHER | ISSUED AND DELAYS |

NOTE: ____________________________________

COMPLETION DATE	DAYS AHEAD OF SCHEDULE	DAYS BEHIND SCHEDULE

SAFETY AND INCIDENTS

SAFETY ISSUES THAT NEED TO BE ADDRESSED	ACCIDENTS / INCIDENTS / STEPS NEEDED TO RESOLVE

SUMMARY OF THE WORK DONE TODAY

IMPORTANT NOTES

NAME	SIGNATURE

TODAY LABOR

INITIALS	TRADE	START	FINISH	PAID HOURS	OVERTIME	COMPANY
☐ EMPLOYEE ☐ CONTRUCTOR		AM	PM			
☐ EMPLOYEE ☐ CONTRUCTOR		AM	PM			
☐ EMPLOYEE ☐ CONTRUCTOR		AM	PM			
☐ EMPLOYEE ☐ CONTRUCTOR		AM	PM			
☐ EMPLOYEE ☐ CONTRUCTOR		AM	PM			
☐ EMPLOYEE ☐ CONTRUCTOR		AM	PM			
☐ EMPLOYEE ☐ CONTRUCTOR		AM	PM			
☐ EMPLOYEE ☐ CONTRUCTOR		AM	PM			

EQUIPMENT ON SITE	NO. OF UNITE	WORKING YES / NO

HIRED EQUIPMENT	NO. OF UNITE	EQUIPMENT RENTED	FROM	RATE

NAME: ______________________ SIGNATURE: ______________________

MO TU WE TH FR SA SU
☐   ☐   ☐   ☐   ☐   ☐   ☐

DATE: ___/___/___

PROJECT:

FOREMAN:

WEATHER

F°_____ C°_____ _____AM _____PM

HOURS DUE TO BAD WEATHER

ISSUED AND DELAYS

NOTE: _______________________

COMPLETION DATE	DAYS AHEAD OF SCHEDULE	DAYS BEHIND SCHEDULE

SAFETY AND INCIDENTS

SAFETY ISSUES THAT NEED TO BE ADDRESSED	ACCIDENTS / INCIDENTS / STEPS NEEDED TO RESOLVE

SUMMARY OF THE WORK DONE TODAY

IMPORTANT NOTES

NAME	SIGNATURE

TODAY LABOR						
INITIALS	TRADE	START	FINISH	PAID HOURS	OVERTIME	COMPANY
☐ EMPLOYEE ☐ CONTRUCTOR		AM	PM			
☐ EMPLOYEE ☐ CONTRUCTOR		AM	PM			
☐ EMPLOYEE ☐ CONTRUCTOR		AM	PM			
☐ EMPLOYEE ☐ CONTRUCTOR		AM	PM			
☐ EMPLOYEE ☐ CONTRUCTOR		AM	PM			
☐ EMPLOYEE ☐ CONTRUCTOR		AM	PM			
☐ EMPLOYEE ☐ CONTRUCTOR		AM	PM			
☐ EMPLOYEE ☐ CONTRUCTOR		AM	PM			

EQUIPMENT ON SITE	NO. OF UNITE	WORKING YES / NO

HIRED EQUIPMENT	NO. OF UNITE	EQUIPMENT RENTED	FROM	RATE

NAME: _______________________ SIGNATURE: _______________________

MO   TU   WE   TH   FR   SA   SU
☐   ☐   ☐   ☐   ☐   ☐   ☐

DATE: ___ / ___ / ___

PROJECT:

FOREMAN:

WEATHER   F° _____   C° _____   _____ AM   _____ PM

HOURS DUE TO BAD WEATHER

ISSUED AND DELAYS

NOTE: ____

COMPLETION DATE	DAYS AHEAD OF SCHEDULE	DAYS BEHIND SCHEDULE

SAFETY AND INCIDENTS

SAFETY ISSUES THAT NEED TO BE ADDRESSED	ACCIDENTS / INCIDENTS / STEPS NEEDED TO RESOLVE

SUMMARY OF THE WORK DONE TODAY

IMPORTANT NOTES

NAME	SIGNATURE

TODAY LABOR

INITIALS	TRADE	START	FINISH	PAID HOURS	OVERTIME	COMPANY
☐ EMPLOYEE ☐ CONTRUCTOR		AM	PM			
☐ EMPLOYEE ☐ CONTRUCTOR		AM	PM			
☐ EMPLOYEE ☐ CONTRUCTOR		AM	PM			
☐ EMPLOYEE ☐ CONTRUCTOR		AM	PM			
☐ EMPLOYEE ☐ CONTRUCTOR		AM	PM			
☐ EMPLOYEE ☐ CONTRUCTOR		AM	PM			
☐ EMPLOYEE ☐ CONTRUCTOR		AM	PM			
☐ EMPLOYEE ☐ CONTRUCTOR		AM	PM			

EQUIPMENT ON SITE	NO. OF UNITE	WORKING YES / NO

HIRED EQUIPMENT	NO. OF UNITE	EQUIPMENT RENTED	FROM	RATE

NAME: _______________________   SIGNATURE: _______________________

MO ☐ TU ☐ WE ☐ TH ☐ FR ☐ SA ☐ SU ☐

DATE: ___/___/___

PROJECT:

FOREMAN:

WEATHER

F° ___ C° ___ ___ AM ___ PM

HOURS DUE TO BAD WEATHER	ISSUED AND DELAYS

NOTE: ____

COMPLETION DATE	DAYS AHEAD OF SCHEDULE	DAYS BEHIND SCHEDULE

SAFETY AND INCIDENTS

SAFETY ISSUES THAT NEED TO BE ADDRESSED	ACCIDENTS / INCIDENTS / STEPS NEEDED TO RESOLVE

SUMMARY OF THE WORK DONE TODAY

IMPORTANT NOTES

NAME	SIGNATURE

TODAY LABOR

INITIALS	TRADE	START	FINISH	PAID HOURS	OVERTIME	COMPANY
☐ EMPLOYEE ☐ CONTRUCTOR		AM	PM			
☐ EMPLOYEE ☐ CONTRUCTOR		AM	PM			
☐ EMPLOYEE ☐ CONTRUCTOR		AM	PM			
☐ EMPLOYEE ☐ CONTRUCTOR		AM	PM			
☐ EMPLOYEE ☐ CONTRUCTOR		AM	PM			
☐ EMPLOYEE ☐ CONTRUCTOR		AM	PM			
☐ EMPLOYEE ☐ CONTRUCTOR		AM	PM			
☐ EMPLOYEE ☐ CONTRUCTOR		AM	PM			

EQUIPMENT ON SITE	NO. OF UNITE	WORKING YES / NO

HIRED EQUIPMENT	NO. OF UNITE	EQUIPMENT RENTED	FROM	RATE

NAME: _______________________   SIGNATURE: _______________________

MO   TU   WE   TH   FR   SA   SU
☐   ☐   ☐   ☐   ☐   ☐   ☐

DATE: /   /

PROJECT:

FOREMAN:

WEATHER   F°_____ C°_____   _____ AM   _____ PM

HOURS DUE TO BAD WEATHER

ISSUED AND DELAYS

NOTE: ____

COMPLETION DATE	DAYS AHEAD OF SCHEDULE	DAYS BEHIND SCHEDULE

SAFETY AND INCIDENTS

SAFETY ISSUES THAT NEED TO BE ADDRESSED	ACCIDENTS / INCIDENTS / STEPS NEEDED TO RESOLVE

SUMMARY OF THE WORK DONE TODAY

IMPORTANT NOTES

NAME	SIGNATURE

TODAY LABOR

INITIALS	TRADE	START	FINISH	PAID HOURS	OVERTIME	COMPANY
☐ EMPLOYEE ☐ CONTRUCTOR		AM	PM			
☐ EMPLOYEE ☐ CONTRUCTOR		AM	PM			
☐ EMPLOYEE ☐ CONTRUCTOR		AM	PM			
☐ EMPLOYEE ☐ CONTRUCTOR		AM	PM			
☐ EMPLOYEE ☐ CONTRUCTOR		AM	PM			
☐ EMPLOYEE ☐ CONTRUCTOR		AM	PM			
☐ EMPLOYEE ☐ CONTRUCTOR		AM	PM			
☐ EMPLOYEE ☐ CONTRUCTOR		AM	PM			

EQUIPMENT ON SITE	NO. OF UNITE	WORKING YES / NO

HIRED EQUIPMENT	NO. OF UNITE	EQUIPMENT RENTED	FROM	RATE

NAME: _______________________________   SIGNATURE: _______________________________

MO   TU   WE   TH   FR   SA   SU
☐   ☐   ☐   ☐   ☐   ☐   ☐

DATE: ___ / ___ / ___

PROJECT:

FOREMAN:

WEATHER

F°_____   C°_____   _____ AM   _____ PM

HOURS DUE TO BAD WEATHER	ISSUED AND DELAYS

NOTE: ____

COMPLETION DATE	DAYS AHEAD OF SCHEDULE	DAYS BEHIND SCHEDULE

SAFETY AND INCIDENTS

SAFETY ISSUES THAT NEED TO BE ADDRESSED	ACCIDENTS / INCIDENTS / STEPS NEEDED TO RESOLVE

SUMMARY OF THE WORK DONE TODAY

IMPORTANT NOTES

NAME	SIGNATURE

INITIALS	TRADE	START	FINISH	PAID HOURS	OVERTIME	COMPANY
☐ EMPLOYEE ☐ CONTRUCTOR		AM	PM			
☐ EMPLOYEE ☐ CONTRUCTOR		AM	PM			
☐ EMPLOYEE ☐ CONTRUCTOR		AM	PM			
☐ EMPLOYEE ☐ CONTRUCTOR		AM	PM			
☐ EMPLOYEE ☐ CONTRUCTOR		AM	PM			
☐ EMPLOYEE ☐ CONTRUCTOR		AM	PM			
☐ EMPLOYEE ☐ CONTRUCTOR		AM	PM			
☐ EMPLOYEE ☐ CONTRUCTOR		AM	PM			

EQUIPMENT ON SITE	NO. OF UNITE	WORKING YES / NO

HIRED EQUIPMENT	NO. OF UNITE	EQUIPMENT RENTED	FROM	RATE

NAME: _________________________   SIGNATURE: _________________________

MO TU WE TH FR SA SU
☐   ☐   ☐   ☐   ☐   ☐   ☐

DATE: /   /

PROJECT:

FOREMAN:

WEATHER

F°____ C°____   ____ AM ____ PM

HOURS DUE TO BAD WEATHER	ISSUED AND DELAYS

NOTE: ____

COMPLETION DATE	DAYS AHEAD OF SCHEDULE	DAYS BEHIND SCHEDULE

SAFETY AND INCIDENTS

SAFETY ISSUES THAT NEED TO BE ADDRESSED	ACCIDENTS / INCIDENTS / STEPS NEEDED TO RESOLVE

SUMMARY OF THE WORK DONE TODAY

IMPORTANT NOTES

NAME	SIGNATURE

TODAY LABOR

INITIALS	TRADE	START	FINISH	PAID HOURS	OVERTIME	COMPANY
☐ EMPLOYEE ☐ CONTRUCTOR		AM	PM			
☐ EMPLOYEE ☐ CONTRUCTOR		AM	PM			
☐ EMPLOYEE ☐ CONTRUCTOR		AM	PM			
☐ EMPLOYEE ☐ CONTRUCTOR		AM	PM			
☐ EMPLOYEE ☐ CONTRUCTOR		AM	PM			
☐ EMPLOYEE ☐ CONTRUCTOR		AM	PM			
☐ EMPLOYEE ☐ CONTRUCTOR		AM	PM			
☐ EMPLOYEE ☐ CONTRUCTOR		AM	PM			

EQUIPMENT ON SITE	NO. OF UNITE	WORKING YES / NO

HIRED EQUIPMENT	NO. OF UNITE	EQUIPMENT RENTED	FROM	RATE

NAME: _______________________ SIGNATURE: _______________________

MO   TU   WE   TH   FR   SA   SU
☐   ☐   ☐   ☐   ☐   ☐   ☐

DATE: /   /

PROJECT:

FOREMAN:

WEATHER

F°____   C°____   ____ AM   ____ PM

HOURS DUE TO BAD WEATHER

ISSUED AND DELAYS

NOTE: ____

COMPLETION DATE	DAYS AHEAD OF SCHEDULE	DAYS BEHIND SCHEDULE

SAFETY AND INCIDENTS

SAFETY ISSUES THAT NEED TO BE ADDRESSED	ACCIDENTS / INCIDENTS / STEPS NEEDED TO RESOLVE

SUMMARY OF THE WORK DONE TODAY

IMPORTANT NOTES

NAME	SIGNATURE

TODAY LABOR

INITIALS	TRADE	START	FINISH	PAID HOURS	OVERTIME	COMPANY
☐ EMPLOYEE ☐ CONTRUCTOR		AM	PM			
☐ EMPLOYEE ☐ CONTRUCTOR		AM	PM			
☐ EMPLOYEE ☐ CONTRUCTOR		AM	PM			
☐ EMPLOYEE ☐ CONTRUCTOR		AM	PM			
☐ EMPLOYEE ☐ CONTRUCTOR		AM	PM			
☐ EMPLOYEE ☐ CONTRUCTOR		AM	PM			
☐ EMPLOYEE ☐ CONTRUCTOR		AM	PM			
☐ EMPLOYEE ☐ CONTRUCTOR		AM	PM			

EQUIPMENT ON SITE	NO. OF UNITE	WORKING YES / NO

HIRED EQUIPMENT	NO. OF UNITE	EQUIPMENT RENTED	FROM	RATE

NAME: ________________________   SIGNATURE: ________________________

MO TU WE TH FR SA SU
☐   ☐   ☐   ☐   ☐   ☐   ☐

DATE: ___ / ___ / ___

PROJECT:

FOREMAN:

WEATHER   F°_____ C°_____ _____ AM _____ PM

HOURS DUE TO BAD WEATHER

ISSUED AND DELAYS

NOTE: ______________________________

COMPLETION DATE	DAYS AHEAD OF SCHEDULE	DAYS BEHIND SCHEDULE

SAFETY AND INCIDENTS

SAFETY ISSUES THAT NEED TO BE ADDRESSED	ACCIDENTS / INCIDENTS / STEPS NEEDED TO RESOLVE

SUMMARY OF THE WORK DONE TODAY

IMPORTANT NOTES

NAME	SIGNATURE

TODAY LABOR

INITIALS	TRADE	START	FINISH	PAID HOURS	OVERTIME	COMPANY
☐ EMPLOYEE ☐ CONTRUCTOR		AM	PM			
☐ EMPLOYEE ☐ CONTRUCTOR		AM	PM			
☐ EMPLOYEE ☐ CONTRUCTOR		AM	PM			
☐ EMPLOYEE ☐ CONTRUCTOR		AM	PM			
☐ EMPLOYEE ☐ CONTRUCTOR		AM	PM			
☐ EMPLOYEE ☐ CONTRUCTOR		AM	PM			
☐ EMPLOYEE ☐ CONTRUCTOR		AM	PM			
☐ EMPLOYEE ☐ CONTRUCTOR		AM	PM			

EQUIPMENT ON SITE	NO. OF UNITE	WORKING YES / NO

HIRED EQUIPMENT	NO. OF UNITE	EQUIPMENT RENTED	FROM	RATE

NAME: _______________________   SIGNATURE: _______________________

MO	TU	WE	TH	FR	SA	SU
☐	☐	☐	☐	☐	☐	☐

DATE: ___ / ___ / ___

PROJECT:

FOREMAN:

WEATHER F°_____ C°_____ _____ AM _____ PM

HOURS DUE TO BAD WEATHER

ISSUED AND DELAYS

NOTE: ___________

COMPLETION DATE	DAYS AHEAD OF SCHEDULE	DAYS BEHIND SCHEDULE

SAFETY AND INCIDENTS

SAFETY ISSUES THAT NEED TO BE ADDRESSED	ACCIDENTS / INCIDENTS / STEPS NEEDED TO RESOLVE

SUMMARY OF THE WORK DONE TODAY

IMPORTANT NOTES

NAME	SIGNATURE

TODAY LABOR

INITIALS	TRADE	START	FINISH	PAID HOURS	OVERTIME	COMPANY
☐ EMPLOYEE ☐ CONTRUCTOR		AM	PM			
☐ EMPLOYEE ☐ CONTRUCTOR		AM	PM			
☐ EMPLOYEE ☐ CONTRUCTOR		AM	PM			
☐ EMPLOYEE ☐ CONTRUCTOR		AM	PM			
☐ EMPLOYEE ☐ CONTRUCTOR		AM	PM			
☐ EMPLOYEE ☐ CONTRUCTOR		AM	PM			
☐ EMPLOYEE ☐ CONTRUCTOR		AM	PM			
☐ EMPLOYEE ☐ CONTRUCTOR		AM	PM			

EQUIPMENT ON SITE	NO. OF UNITE	WORKING YES / NO

HIRED EQUIPMENT	NO. OF UNITE	EQUIPMENT RENTED	FROM	RATE

NAME: _______________________

SIGNATURE: _______________________

MO TU WE TH FR SA SU
☐   ☐   ☐   ☐   ☐   ☐   ☐

DATE: ____ / ____ / ____

PROJECT:

FOREMAN:

WEATHER

F° ____ C° ____ ____ AM ____ PM

HOURS DUE TO BAD WEATHER

ISSUED AND DELAYS

NOTE: ____

COMPLETION DATE	DAYS AHEAD OF SCHEDULE	DAYS BEHIND SCHEDULE

SAFETY AND INCIDENTS

SAFETY ISSUES THAT NEED TO BE ADDRESSED	ACCIDENTS / INCIDENTS / STEPS NEEDED TO RESOLVE

SUMMARY OF THE WORK DONE TODAY

IMPORTANT NOTES

NAME	SIGNATURE

TODAY LABOR

INITIALS	TRADE	START	FINISH	PAID HOURS	OVERTIME	COMPANY
☐ EMPLOYEE ☐ CONTRUCTOR		AM	PM			
☐ EMPLOYEE ☐ CONTRUCTOR		AM	PM			
☐ EMPLOYEE ☐ CONTRUCTOR		AM	PM			
☐ EMPLOYEE ☐ CONTRUCTOR		AM	PM			
☐ EMPLOYEE ☐ CONTRUCTOR		AM	PM			
☐ EMPLOYEE ☐ CONTRUCTOR		AM	PM			
☐ EMPLOYEE ☐ CONTRUCTOR		AM	PM			
☐ EMPLOYEE ☐ CONTRUCTOR		AM	PM			

EQUIPMENT ON SITE	NO. OF UNITE	WORKING YES / NO

HIRED EQUIPMENT	NO. OF UNITE	EQUIPMENT RENTED	FROM	RATE

NAME: _______________________   SIGNATURE: _______________________

MO   TU   WE   TH   FR   SA   SU   DATE: ____ / ____ / ____

PROJECT: **FOREMAN:**

WEATHER   F° ____   C° ____   ____ AM   ____ PM   **HOURS DUE TO BAD WEATHER**   **ISSUED AND DELAYS**

NOTE: ____

COMPLETION DATE	DAYS AHEAD OF SCHEDULE	DAYS BEHIND SCHEDULE

SAFETY AND INCIDENTS

SAFETY ISSUES THAT NEED TO BE ADDRESSED	ACCIDENTS / INCIDENTS / STEPS NEEDED TO RESOLVE

SUMMARY OF THE WORK DONE TODAY

IMPORTANT NOTES

NAME	SIGNATURE

TODAY LABOR

INITIALS	TRADE	START	FINISH	PAID HOURS	OVERTIME	COMPANY
☐ EMPLOYEE ☐ CONTRUCTOR		AM	PM			
☐ EMPLOYEE ☐ CONTRUCTOR		AM	PM			
☐ EMPLOYEE ☐ CONTRUCTOR		AM	PM			
☐ EMPLOYEE ☐ CONTRUCTOR		AM	PM			
☐ EMPLOYEE ☐ CONTRUCTOR		AM	PM			
☐ EMPLOYEE ☐ CONTRUCTOR		AM	PM			
☐ EMPLOYEE ☐ CONTRUCTOR		AM	PM			
☐ EMPLOYEE ☐ CONTRUCTOR		AM	PM			

EQUIPMENT ON SITE	NO. OF UNITE	WORKING YES / NO

HIRED EQUIPMENT	NO. OF UNITE	EQUIPMENT RENTED	FROM	RATE

NAME: ______________________   SIGNATURE: ______________________

MO ☐　TU ☐　WE ☐　TH ☐　FR ☐　SA ☐　SU ☐

DATE: ____ / ____ / ____

PROJECT:

FOREMAN:

WEATHER ☁️ ⛅ ☁️ ❄️ ☀️ 🌧️ ⛈️

F°_____　C°_____ _____ AM _____ PM

| HOURS DUE TO BAD WEATHER | ISSUED AND DELAYS |

NOTE: ____

COMPLETION DATE	DAYS AHEAD OF SCHEDULE	DAYS BEHIND SCHEDULE

SAFETY AND INCIDENTS

SAFETY ISSUES THAT NEED TO BE ADDRESSED	ACCIDENTS / INCIDENTS / STEPS NEEDED TO RESOLVE

SUMMARY OF THE WORK DONE TODAY

IMPORTANT NOTES

NAME	SIGNATURE

TODAY LABOR

INITIALS	TRADE	START	FINISH	PAID HOURS	OVERTIME	COMPANY
☐ EMPLOYEE ☐ CONTRUCTOR		AM	PM			
☐ EMPLOYEE ☐ CONTRUCTOR		AM	PM			
☐ EMPLOYEE ☐ CONTRUCTOR		AM	PM			
☐ EMPLOYEE ☐ CONTRUCTOR		AM	PM			
☐ EMPLOYEE ☐ CONTRUCTOR		AM	PM			
☐ EMPLOYEE ☐ CONTRUCTOR		AM	PM			
☐ EMPLOYEE ☐ CONTRUCTOR		AM	PM			
☐ EMPLOYEE ☐ CONTRUCTOR		AM	PM			

EQUIPMENT ON SITE	NO. OF UNITE	WORKING YES / NO

HIRED EQUIPMENT	NO. OF UNITE	EQUIPMENT RENTED	FROM	RATE

NAME: _______________________   SIGNATURE: _______________________

| MO | TU | WE | TH | FR | SA | SU | | DATE: ___ / ___ / ___ |
| ☐ | ☐ | ☐ | ☐ | ☐ | ☐ | ☐ | | |

PROJECT:

FOREMAN:

WEATHER

F° ___ C° ___ ___ AM ___ PM

| HOURS DUE TO BAD WEATHER | ISSUED AND DELAYS |

NOTE: ____

COMPLETION DATE	DAYS AHEAD OF SCHEDULE	DAYS BEHIND SCHEDULE

SAFETY AND INCIDENTS

SAFETY ISSUES THAT NEED TO BE ADDRESSED	ACCIDENTS / INCIDENTS / STEPS NEEDED TO RESOLVE

SUMMARY OF THE WORK DONE TODAY

IMPORTANT NOTES

NAME	SIGNATURE

TODAY LABOR

INITIALS	TRADE	START	FINISH	PAID HOURS	OVERTIME	COMPANY
☐ EMPLOYEE ☐ CONTRUCTOR		AM	PM			
☐ EMPLOYEE ☐ CONTRUCTOR		AM	PM			
☐ EMPLOYEE ☐ CONTRUCTOR		AM	PM			
☐ EMPLOYEE ☐ CONTRUCTOR		AM	PM			
☐ EMPLOYEE ☐ CONTRUCTOR		AM	PM			
☐ EMPLOYEE ☐ CONTRUCTOR		AM	PM			
☐ EMPLOYEE ☐ CONTRUCTOR		AM	PM			
☐ EMPLOYEE ☐ CONTRUCTOR		AM	PM			

EQUIPMENT ON SITE	NO. OF UNITE	WORKING YES / NO

HIRED EQUIPMENT	NO. OF UNITE	EQUIPMENT RENTED	FROM	RATE

NAME: _________________________   SIGNATURE: _________________________

MO   TU   WE   TH   FR   SA   SU
☐   ☐   ☐   ☐   ☐   ☐   ☐

DATE: /   /

PROJECT:

FOREMAN:

WEATHER

F°______   C°______   ______ AM   ______ PM

HOURS DUE TO
BAD WEATHER

ISSUED AND DELAYS

NOTE: ____

COMPLETION DATE	DAYS AHEAD OF SCHEDULE	DAYS BEHIND SCHEDULE

SAFETY AND INCIDENTS

SAFETY ISSUES THAT NEED TO BE ADDRESSED	ACCIDENTS / INCIDENTS / STEPS NEEDED TO RESOLVE

SUMMARY OF THE WORK DONE TODAY

IMPORTANT NOTES

NAME	SIGNATURE

TODAY LABOR

INITIALS	TRADE	START	FINISH	PAID HOURS	OVERTIME	COMPANY
☐ EMPLOYEE ☐ CONTRUCTOR		AM	PM			
☐ EMPLOYEE ☐ CONTRUCTOR		AM	PM			
☐ EMPLOYEE ☐ CONTRUCTOR		AM	PM			
☐ EMPLOYEE ☐ CONTRUCTOR		AM	PM			
☐ EMPLOYEE ☐ CONTRUCTOR		AM	PM			
☐ EMPLOYEE ☐ CONTRUCTOR		AM	PM			
☐ EMPLOYEE ☐ CONTRUCTOR		AM	PM			
☐ EMPLOYEE ☐ CONTRUCTOR		AM	PM			

EQUIPMENT ON SITE	NO. OF UNITE	WORKING YES / NO

HIRED EQUIPMENT	NO. OF UNITE	EQUIPMENT RENTED	FROM	RATE

NAME: _______________________________   SIGNATURE: _______________________________

MO ☐ TU ☐ WE ☐ TH ☐ FR ☐ SA ☐ SU ☐

DATE: ___ / ___ / ___

PROJECT: __________________

FOREMAN: __________________

WEATHER

F° ____   C° ____   ____ AM   ____ PM

HOURS DUE TO BAD WEATHER	ISSUED AND DELAYS

NOTE: __________________

COMPLETION DATE	DAYS AHEAD OF SCHEDULE	DAYS BEHIND SCHEDULE

SAFETY AND INCIDENTS

SAFETY ISSUES THAT NEED TO BE ADDRESSED	ACCIDENTS / INCIDENTS / STEPS NEEDED TO RESOLVE

SUMMARY OF THE WORK DONE TODAY

IMPORTANT NOTES

NAME	SIGNATURE

TODAY LABOR

INITIALS	TRADE	START	FINISH	PAID HOURS	OVERTIME	COMPANY
☐ EMPLOYEE ☐ CONTRUCTOR		AM	PM			
☐ EMPLOYEE ☐ CONTRUCTOR		AM	PM			
☐ EMPLOYEE ☐ CONTRUCTOR		AM	PM			
☐ EMPLOYEE ☐ CONTRUCTOR		AM	PM			
☐ EMPLOYEE ☐ CONTRUCTOR		AM	PM			
☐ EMPLOYEE ☐ CONTRUCTOR		AM	PM			
☐ EMPLOYEE ☐ CONTRUCTOR		AM	PM			
☐ EMPLOYEE ☐ CONTRUCTOR		AM	PM			

EQUIPMENT ON SITE	NO. OF UNITE	WORKING YES / NO

HIRED EQUIPMENT	NO. OF UNITE	EQUIPMENT RENTED	FROM	RATE

NAME: _______________________   SIGNATURE: _______________________

MO TU WE TH FR SA SU
☐   ☐   ☐   ☐   ☐   ☐   ☐

DATE: ___ / ___ / ___

PROJECT:

FOREMAN:

WEATHER

F°_____ C°_____   _____ AM _____ PM

HOURS DUE TO BAD WEATHER	ISSUED AND DELAYS

NOTE: ____

COMPLETION DATE	DAYS AHEAD OF SCHEDULE	DAYS BEHIND SCHEDULE

SAFETY AND INCIDENTS

SAFETY ISSUES THAT NEED TO BE ADDRESSED	ACCIDENTS / INCIDENTS / STEPS NEEDED TO RESOLVE

SUMMARY OF THE WORK DONE TODAY

IMPORTANT NOTES

NAME	SIGNATURE

TODAY LABOR

INITIALS	TRADE	START	FINISH	PAID HOURS	OVERTIME	COMPANY
☐ EMPLOYEE ☐ CONTRUCTOR		AM	PM			
☐ EMPLOYEE ☐ CONTRUCTOR		AM	PM			
☐ EMPLOYEE ☐ CONTRUCTOR		AM	PM			
☐ EMPLOYEE ☐ CONTRUCTOR		AM	PM			
☐ EMPLOYEE ☐ CONTRUCTOR		AM	PM			
☐ EMPLOYEE ☐ CONTRUCTOR		AM	PM			
☐ EMPLOYEE ☐ CONTRUCTOR		AM	PM			
☐ EMPLOYEE ☐ CONTRUCTOR		AM	PM			

EQUIPMENT ON SITE	NO. OF UNITE	WORKING YES / NO

HIRED EQUIPMENT	NO. OF UNITE	EQUIPMENT RENTED	FROM	RATE

NAME: _______________________   SIGNATURE: _______________________

MO   TU   WE   TH   FR   SA   SU
☐   ☐   ☐   ☐   ☐   ☐   ☐

DATE: /   /

PROJECT:

FOREMAN:

WEATHER

F°_____   C°_____   _____AM   _____PM

HOURS DUE TO BAD WEATHER

ISSUED AND DELAYS

NOTE: _______________________

COMPLETION DATE	DAYS AHEAD OF SCHEDULE	DAYS BEHIND SCHEDULE

SAFETY AND INCIDENTS

SAFETY ISSUES THAT NEED TO BE ADDRESSED	ACCIDENTS / INCIDENTS / STEPS NEEDED TO RESOLVE

SUMMARY OF THE WORK DONE TODAY

IMPORTANT NOTES

NAME	SIGNATURE

TODAY LABOR

INITIALS	TRADE	START	FINISH	PAID HOURS	OVERTIME	COMPANY
☐ EMPLOYEE ☐ CONTRUCTOR		AM	PM			
☐ EMPLOYEE ☐ CONTRUCTOR		AM	PM			
☐ EMPLOYEE ☐ CONTRUCTOR		AM	PM			
☐ EMPLOYEE ☐ CONTRUCTOR		AM	PM			
☐ EMPLOYEE ☐ CONTRUCTOR		AM	PM			
☐ EMPLOYEE ☐ CONTRUCTOR		AM	PM			
☐ EMPLOYEE ☐ CONTRUCTOR		AM	PM			
☐ EMPLOYEE ☐ CONTRUCTOR		AM	PM			

EQUIPMENT ON SITE	NO. OF UNITE	WORKING YES / NO

HIRED EQUIPMENT	NO. OF UNITE	EQUIPMENT RENTED	FROM	RATE

NAME: ______________________________   SIGNATURE: ______________________________

IMPORTANT TELEPHONE NUMBER

▶ NAME ______________________________ PHONE ______________________________
    EMAIL ____

▶ NAME ______________________________ PHONE ______________________________
    EMAIL ____

▶ NAME ______________________________ PHONE ______________________________
    EMAIL ____

▶ NAME ______________________________ PHONE ______________________________
    EMAIL ____

▶ NAME ______________________________ PHONE ______________________________
    EMAIL ____

▶ NAME ______________________________ PHONE ______________________________
    EMAIL ____

▶ NAME ______________________________ PHONE ______________________________
    EMAIL ____

▶ NAME ______________________________ PHONE ______________________________
    EMAIL ____

▶ NAME ______________________________ PHONE ______________________________
    EMAIL ____

▶ NAME ______________________________ PHONE ______________________________
    EMAIL ____

▶ NAME ______________________________ PHONE ______________________________
    EMAIL ____

▶ NAME ______________________________ PHONE ______________________________
    EMAIL ____

▶ NAME ______________________________ PHONE ______________________________
    EMAIL ____

▶ NAME ______________________________ PHONE ______________________________
    EMAIL ____